Mastering Split-Range Control

To Improve Process Control Results

George Buckbee, P.E.

PIDTutor offers many resources to help you to develop your process control skills. Visit the following sites on-line:

Main Web Site: http://www.pidtutor.com

Blog: http://wordpress.com/processcontrolguru

This document is for educational purposes only. There are many factors affecting control system design and implementation. Control system design and implementation should be accomplished by trained professionals. PIDTutor assumes no liability for the implementation of control systems as described in this document.

About the Author

George Buckbee, P.E.is the founder and president of PIDTutor. He has over 20 years' experience in the practical application of process control. George holds a B.S. and M.S. in Chemical Engineering, and is a registered professional engineer.

Table of Contents

Introduction to Split-Range Control

Split-range control is a powerful control strategy that uses two or more valves to control a single process variable. Figure 1 shows a typical split-range control system, where two valves are used to control the temperature of a jacketed tank. One valve supplies heating fluid, and the other supplies cooling fluid to the jacket of the tank.

The single temperature controller (TIC) adjusts its output up and down to control temperature. The relay, labeled (Y), passes on the control signal to open one of the two valves. When the control output is at:

 0%, the cooling fluid valve will be fully open, and the heating fluid valve is closed.

25%, the cooling fluid valve is half-open. The heating valve remains closed.

50%, both valves are fully closed.

75%, the heating valve is half-open. The cooling fluid valve remains closed.

100%, the heating fluid valve is fully open. The cooling fluid valve remains closed.

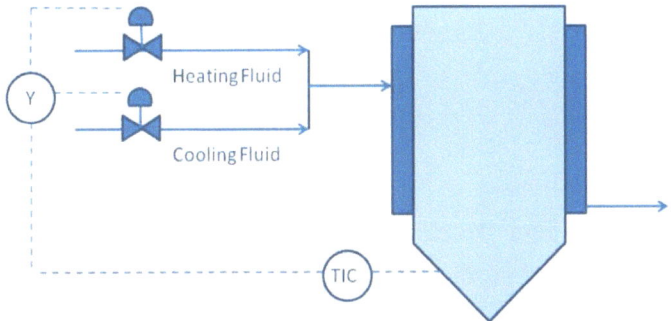

Figure 1. A Typical Split-Range Control System

In this way, a single controller is used to maintain a single temperature. The arrangement is called "split-range" because the output range is split into two parts. Half of the output range is used to open or close the cooling valve, and half is used to open or close the heating valve.

What You Will Learn:

In this book, you will learn about all aspects of split-range control, including:

- How split-range control works
- The costs and benefits of split-range control
- When to choose split-range as a control strategy
- When NOT to use a split-range strategy
- How to implement split-range control in a:
    - DCS
    - PLC
    - Pneumatic system
- How to calibrate valves in a split-range system
- Commissioning split-range systems
- Tuning methods for split-range controls controllers
- How to troubleshoot split-range controllers

How Split-Range Control Works

Basics

A split-range controller is used when a single control valve is not enough to control the process variable (PV). For example, if both heating and cooling are required to control a temperature, then you may need a split-range controller. One valve controls the heating medium, such as steam or heating fluid, and the other controls the cooling medium, such as chilled water, as shown in Figure 1.

The controller, in fact, does not "know" that it is being used in a split-range scenario. The controller simply adjusts its output up or down to try to maintain the process variable. The controller most often is in the form of a PI or PID controller. However, the split-range approach is equally applicable to other controller types, such as fuzzy-logic controllers.

The key components of a split-range control scheme are:

- A single process measurement (Temperature, pressure, pH, level, etc.)

- A single controller

- A signal splitter of some sort, as shown by the symbol (Y) in Figure 1. This may be a hardware or software device, as we'll see later.

- Two or more valves (or other actuated devices), each controlling a different supplied medium.

The signal splitter follows a simple curve to determine which of the valves should be open, and by how much, based upon the controller's output signal. Figure 2 shows the valve positions for a typical heating/cooling application.

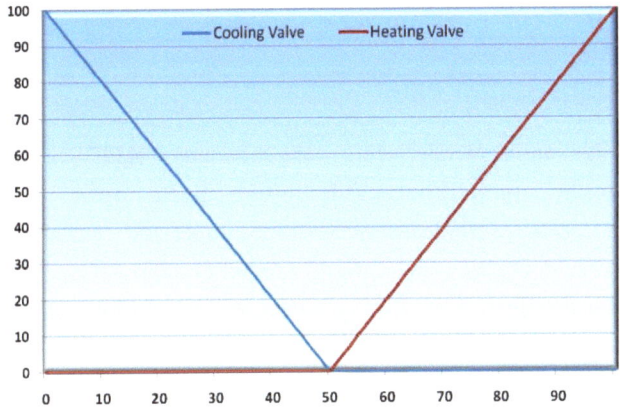

Figure 2. Valve Positions for Split-Range Heating/Cooling Application

The Costs and Benefits of Split-Range Control

Costs

A split-range controller costs more than a typical PID controller, primarily because it has two valves. The relay/signal splitter is a small cost by comparison. A slight increase in engineering, design, installation, and commissioning effort will also be associated with split-range controllers. The extent of the additional cost varies widely with the cost of the valve.

Benefits

The primary benefit of split-range control is that you can use a single control loop for a single process variable. This simplifies the control in several ways:

> There is a single operator interface.
> There is a single instrument, one set of I/O, and one controller.
> There is only one loop to tune.
> The control will automatically switch between valves, with no extra programming logic, or extra effort on the part of the operator.

In effect, split-range controls allow a smooth transition, and single controller across two different operating regimes.

When to Choose Split-Range as a Control Strategy

Certain process situations lend themselves directly to split-range control strategies. In other scenarios, however, split-range may be less effective than other methods. In this section, we explore some applications of split-range control.

Typical Split-Range Applications

Split-range control is appropriate for scenarios where more than one valve (or stream) is needed to control the process variable (PV). Some typical examples include:

> Temperature control with a combination of heating and cooling media.
> pH Control with acid and base reagents.
> Bleed of excess gas pressure to a flare or other exhaust.

In each case, you can see that there are actually two distinct flows being used to adjust the process variable.

Batch Reactor Scenario

A classic example of split-range control is a jacketed batch reactor. The jacket may be filled with hot or cold solution, such as water, glycol, or similar medium. Hot and cold solutions each have their own delivery system. A diagram of this system is shown in Figure 1.

In a typical scenario, the reactor temperature follows a pre-defined path through time. First, the reactor must be heated. The hot solution is used to bring the reactor up to temperature. At some point, a chemical reaction starts in the reactor, giving off heat. The hot solution valve starts to close. As the reaction continues, cooling is needed to hold the temperature constant. The hot solution valve closes, and the cooling valve opens. At the end of the batch, the reactor must be cooled again. Figure 3 shows the temperature profile for this batch, as well as the valve positions for the heating and cooling valves.

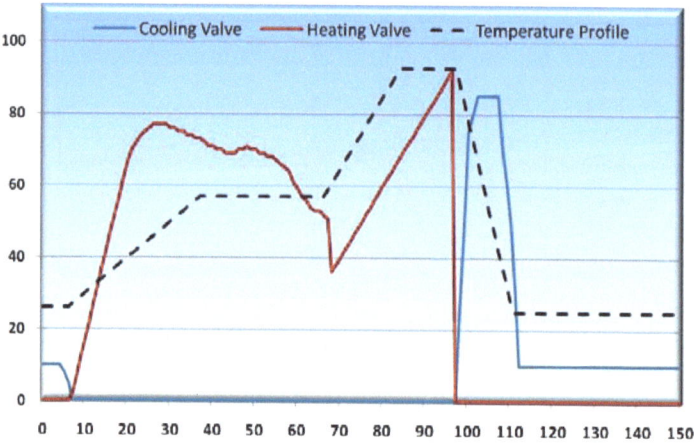

Figure 3. Example Reactor Profile and Split-Range Valve Positions

Bleed & Vent Scenario

In another scenario pressure control is normally maintained using a small valve for gas pressure control.   The valuable gas is used downstream in the process. Under some process conditions, a large amount of gas is produced, and this excess gas must be vented to keep the process under control. A larger "vent valve" may be used to extend the range of the process, as shown in Figure 4. The valve positions for the normal and vent valves are shown in Figure 5.

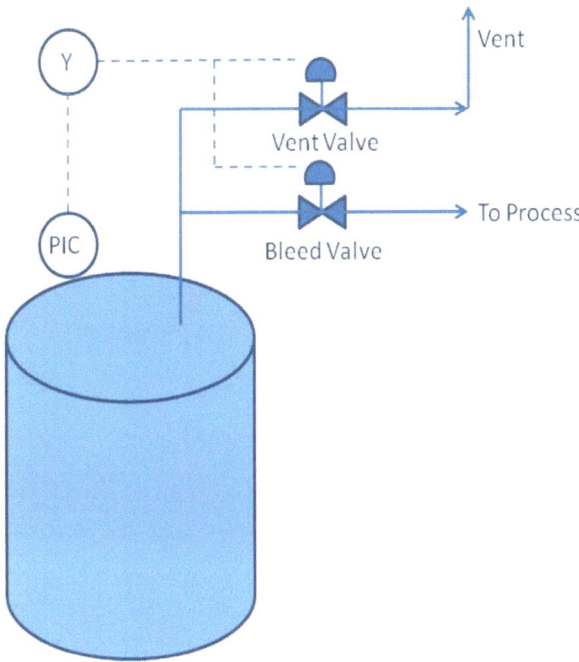

Figure 4. Pressure Control with Bleed and Vent Valves

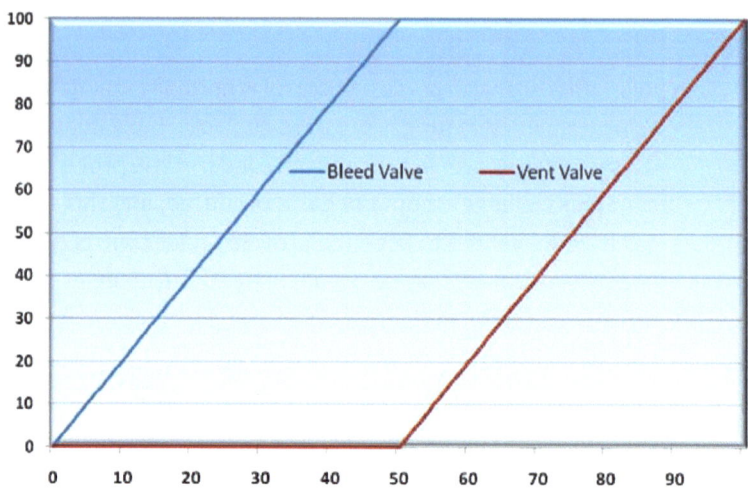

Figure 5. Bleed and Vent Valve Positions

Characteristics of a Good Split-Range Application

Not every scenario lends itself to split-range control. There are several criteria that define a good split-range application:

- A single process variable is being controlled.

- A single valve cannot be used because:

- There are multiple media (heating/cooling, acid/base)

- OR the process stream must be split (bleed/vent)

- It is important to have smooth, continuous control over the entire range.

- A single controller is preferred, to simplify the operator interface.

When NOT to Choose a Split-Range Strategy

Split-range controllers are often mis-applied to extend turndown, as in the case of high-flow low-flow requirements. In this case, you have two valves, one large and one small, both carrying the same fluid, and used together to control the same PV. While, on the surface, this seems a good idea, a much better control strategy, known as Mid-Ranging Control, can be applied here.

How to Implement Cascade Control in a Pneumatic System

In traditional pneumatic systems, split-range controls were configured mechanically, through the use of a pneumatic relay and/or valve positioned calibrations. Figure 6 illustrates this type of system.

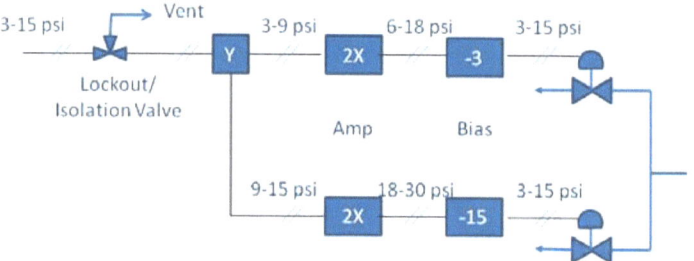

Figure 6. A Pneumatic Split-Range System

This design is relatively simple to implement and maintain. It can be applied in most any scenario with pneumatic-actuated valves. Each valve must have a positioned, calibrated from 0 to 100% opening for the expected range of air pressure (3-9psi or 9-15psi).

However, the relay technique is not recommended for good control. The reason is that, during some operating conditions, the relay will tend to cycle between the two valves. This creates 2 problems:

1. Process cycling
2. Excessive wear-and-tear on the valve.

A further disadvantage is that the accuracy of the control is essentially cut in two, as you are dependent on the pneumatic accuracy of the positioner.

A refined approach, using two "Gain plus Bias" relays, provides more accurate control signals to each valve, but increases the cost, complexity, and maintenance. When precise control is needed, this technique is preferred. Note that the cycling effects can be reduced by overlapping the control ranges of these two relays. This is explained later, in the section on Valve Calibration.

How to Implement Split-Range Control in a DCS

In a modern Distributed Control System (DCS), you can achieve accuracy and precise control through proper configuration of a split-range controller. Figure 7 shows a typical implementation in a DCS.

The advantages of a DCS application are that you can range the output blocks so that each valve will see its full range of actuation (4-20 mA, or 3-15 psi), providing the full accuracy of the valve. Also, the valves can be calibrated like every other valve in the plant, with 4-20 mA representing 0-100% (or 100%-0%) of position.

A DCS also allows for the simple implementation of a characterizer, which is discussed below, in the section on tuning split-range controls.

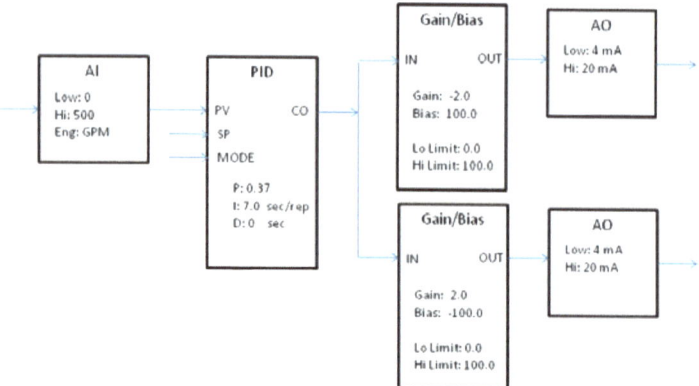

Figure 7. DCS Configuration of Split-Range Control

How to Implement Split-Range Control in a PLC

Implementation in a PLC is very similar to implementation in a DCS, although you may have to custom-develop the splitter module.

In a PLC, the simplest form of the splitter module is to use two X-Y Lookup Tables. One table is used for each control valve. This has the added advantage that you can compensate for non-linearities using the same arrangement. Compensation for non-linearity is discussed later.

Alternately, you may choose to use equations to implement the splitter.

How to Choose Split Range Control Options

Valve Sizing

The valves must be sized according to the process needs. It is a mistake to assume that both valves should be the same size. For example, in the batch reactor example used above, the heating valve will be sized large enough to heat the vessel during the heat-up phase. The cooling valve must be large enough to:

- Remove the heat of reaction during the reaction phase

- Quickly cool the reactor after reaction is complete.

The larger of these two requirements will determine the valve size.

Valve Selection

The type of control valve depends primarily on the fluids to be handled. Select each control valve individually, according to the fluid type, flow requirements, temperatures, corrosion-resistance, and pressure-drop.

As secondary considerations only, the following factors are important:

Linear installed characteristics will simplify the valve tuning. This is covered in more detail below, in the section "Tuning Split-Range Controllers"

If you have a choice, the system should be designed so that the effective process gain of each valve is roughly the same. That is, if you are controlling temperature, the effect of a 1% valve movement by either valve will cause the temperature to move by roughly the same amount. Again, there is more detail on this in the section "Tuning Split-Range Controllers"

Positioners

As a general rule, split-range control valves should include positioners. A positioner helps to ensure that the control valve is actually delivering the valve stem position required by the split-range controller. This is especially important during the transition from one valve to the other.

There is one major exception to this rule: If the controlled PV is a very fast-reacting loop, such as flow, then positioners are not recommended.

Commissioning Split-Range Systems

Valve Calibration

Calibration of split-range valves has a direct impact on the performance of the system. There are 2 schools of thought on calibration of split-range valves. Method 1 allows for precise control, while Method 2 minimizes resource consumption at the transition.

Method 1: Precise Control Required.

It is important to recognize that many control valves do not actually lift off of their seats until the valve stem position is actually 2 or 3% of scale. Without an overlap, it is very likely that <u>both</u> valves will be fully closed for several percent of range during the transition from one valve to the other. During this time, there is <u>no control</u> of the PV. Very typically, this causes oscillation between the two valves, as the control cycles through this "dead area", then back again.

By overlapping the calibration curves, we can ensure that there is a smooth transition between valves. . When precise control of the PV is required, then the valves should be calibrated to allow for overlap, as shown in Figure 8.

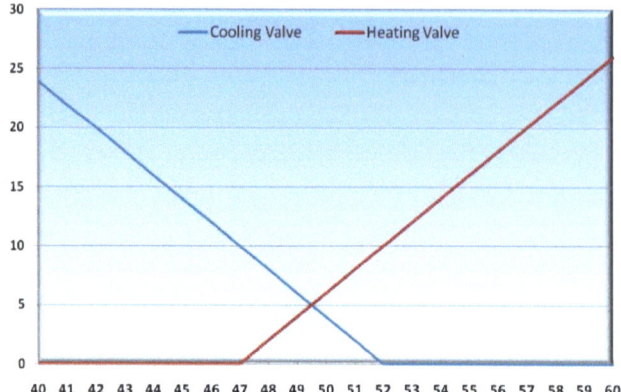

Figure 8. Valve Calibration Overlap, For Precise Control

The downside of this approach is that, during the transition, <u>both</u> valves will actually be open. In a temperature control scenario, that means you will be using heating and cooling mediums <u>at the same time</u>. This inefficiency is the price you pay for better control during the transition.

For bench calibration using the precise control method, Table 1 contains a set of recommendations for valve stem positions, given various air pressures or mA signal values.

Method 2: Minimize Resource Consumption.

Because the precise control method wastes resources, some people prefer the "gap method" for calibration of split-range valves. In this method, an intentional gap is left between the valve curves, as shown in Figure 9.

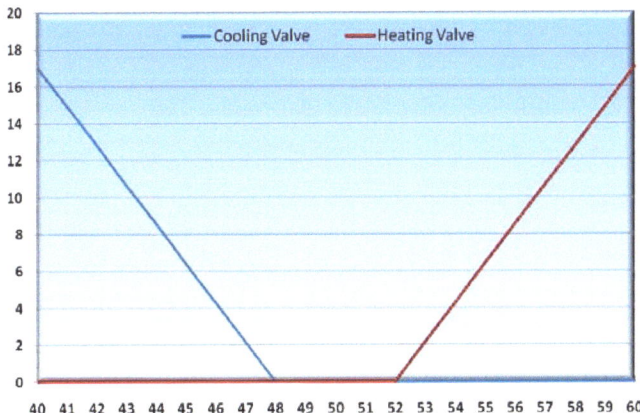

Figure 9. Valve Calibration Gap, To Minimize Resource Consumption

To avoid continuous cycling during the valve transition, "Gap" or "deadband" control algorithms are often recommended. A typical Gap algorithm will force no control action until the controller error (PV – SP) exceeds some limit, usually 1% or more of PV span. While this method does not actually eliminate cycling, it does reduce the frequency of the cycles.

For bench calibration using the "minimize resource consumption" method, Table 2 contains a set of recommendations for valve stem positions at a set of air pressures or mA signal values.

Operator Training

Of course, operators must be trained in split-range valve applications. It can be very confusing to see a control output at 75%, while the actual valve is 50% open. Operators should be trained in the following aspects:

- Safety and lockout procedures may require locking out two valves, rather than one.

- Understanding that the controller output signal will not equal the valve stem position.
- Understanding that, due to the nature of a PID controller, either valve may be open when the PV is above (or below) the setpoint. It is a common misconception that both valve should be closed when the PV equals the setpoint.

Tuning Split-Range Controllers

Tuning a single controller with two or more valves can pose some problems. The primary issue, of course, is that the process response will be highly non-linear. Figure 10 shows both a linear valve response and the typical highly non-linear response of a split-range system. There are 3 major factors that cause non-linear responses in split-range systems:

Two different valves, with two different processes are being used. For example, the temperature response to a 10% increase in steam valve position is unlikely to be similar to a 10% decrease in cooling water valve position.

Both valves operate over a significant portion of their ranges, including operation near the low end of valve position, from 0% to 10% open. Most control valves are very non-linear in this range.

The transition from one valve to the other is not smooth. This is especially true if you have chosen the "Minimize Resource Consumption" calibration option explained above.

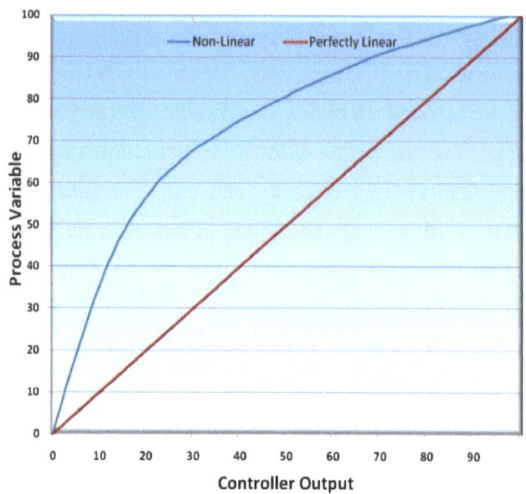

Figure 10. Linear and Non-Linear Valve Responses

You can estimate the degree of non-linearity by plotting the Process Variable (PV) against the controller output (CO) over the full range of the controller, as shown in Figure 11. Compare the slope of the line at its steepest point to the slope at its most shallow point. If the steepest slope is greater than 3 times the shallow slope, then linearization is recommended. If the ratio if the slopes is less than 3, then the loop can be tuned without linearization, using a robust tuning approach.

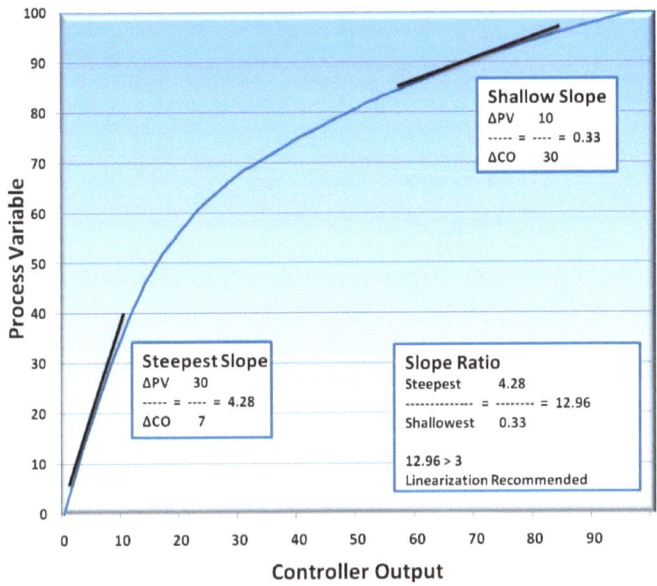

Figure 11. Linear and Non-Linear Valve Responses

Linearization

In "the old days", linearization was accomplished with a mechanical cam device. The shape of the cam, driving a valve positioner, would offset the valve non-linearity. The cam was often custom-designed or modified so that the resulting PV-versus-CO curve was relatively linear, drawing a straight line.

In today's modern control system, linearization is accomplished in software, using a "characterizer". The characterizer may take the form of an equation or a simple X-Y lookup table. Figure 12 shows how a characterizer works in concert with the PID Loop to control the valves.

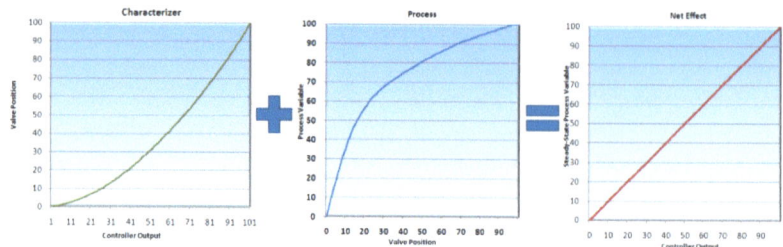

Figure 12. How a Characterizer Offsets Process Non-Linearity

To design a characterizer, step the process through the entire range of valve operation. Since normal operating conditions may not allow this, the step testing should be scheduled during a start-up or shut-down. Take the resulting steady-state PV-versus-CO curve, and invert it to create an X-Y lookup table. Note that automated software solutions are available to accomplish this, as shown in Figure 13.

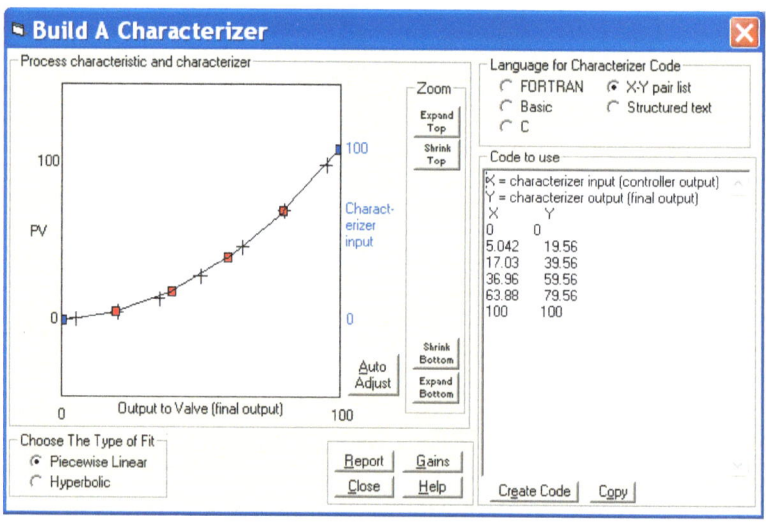

Figure 13. Software Tools To Design a Characterizer

Tuning

From the point of view of the controller, the characterizer, signal splitter, and valves are all part of the overall process. Keep in mind that changes to any of these components will likely require new controller tuning.

A structured approach to tuning is recommended. Most structured approaches require bump testing to determine a process model. When tuning split-range systems, bump tests should include:

Several bumps, throughout the range of operation for both valves.

Steps in both the upward and downward directions.

Develop the process models for each bump test, making special note of any irregularities. Use the most conservative (worst case) process model to determine the controller tuning. The worst case is the one with the:

- Largest Dead-time

- Largest Process Gain

- Smallest Lag Time

Choose "robust" tuning for a split-range controller. The term "robust" means that the controller tuning will perform well under a variety of conditions. In practical terms, this means that you should not choose aggressive tuning methods, such as Ziegler-Nichols quarter-amplitude damping.

Gain Scheduling

Another approach to linearity problems is to apply "Gain Scheduling". With this approach, two or more sets of tuning are applied to the loop. A typical arrangement would be one set of tuning for one valve, another set of tuning for the other. Simple logic in the control system determines which set of tuning should apply.

No Integral Near Transition

Another approach to split-range control tuning is to eliminate the integral action while near the transition zone between the two valves. This is a programmatic way to eliminate the cycling that often occurs as you transition from one valve to the other. Of course, this will result in a loss of control precision during that transition. Typically, integral action is eliminated within 1% either side of the transition.

Troubleshooting Split-Range Controllers

A great strength of split-range controllers is that they behave much like other PID controllers. There are, however, some special problems that are unique to split-range controllers. These problems, their causes and corrective actions are shown below in the "Trouble-Cause-Correction" table.

Trouble	Cause	Correction
PV cycles between two values near the valve transition.	Integral action drives the loop back and forth across the transition.	Overlap valve calibration curves, as shown in Figure 5. Or, eliminate integral action when near the transition.
Poor control with one valve, Ok with the other.	Size or dynamic behavior of valves are not matched.	Consider the use of a characterizer or use Gain Scheduling.
Cannot "shut off" the loop. Fluid leaks by at 50% position.	Calibration for precise control may not allow both valves 100% closed.	Consider a relay to dump air supply when tight shutoff is needed. Or, use the "minimize resources" approach to valve calibration.
Tuning OK in one area, bad in another.	Non-linear valve response.	Consider alternate valve trim, or characterizer.
Valve position does not match control output.	It's not supposed to!	Training. Ensure that operators understand how the split-range system operates.

Conclusions

The split-range control scheme is a useful tool when controlling a single process variable with two or more valves. Split-range offers the following advantages:

- It behaves like a single control loop.

- There is automatic switching between the valves.

- It offers options for precise control or to minimize resources.

Proper control performance of the split-range system depends upon proper configuration, valve calibration, tuning, and operator training.

Use the techniques described above to ensure that each split-range application operates effectively.

References & Further Reading

Split Range Control Methods, Bela Liptak et al, *Control* magazine, 2005
http://www.controlglobal.com/articles/2005/580.html

Greene, John. Split Range Control, Contek,

http://www.contek-systems.co.uk/Documents/split_range_control.htm

Liptak, Bela, The Instrument Engineer's Handbook, "Process Control", Chapter 4.3 Accessories & Positioners, Chilton, 1995., Radnor, PA.

Figure 13 Provided courtesy of ExperTune, Inc.